ELECTRICITY AND MAGNETISM

Chris Oxlade

An imprint of Hodder Children's Books

Titles in the *Science Files* series:

Electricity and Magnetism • Forces and Motion • Light and Sound • The Solar System

Science Files is a simplified and updated version of Hodder Wayland's *Science Fact Files.*

For more information on this series and other Hodder Wayland titles, go to www.hodderwayland.co.uk

Editor: Katie Sergeant
Designer: Simon Borrough
Typesetter: Victoria Webb
Illustrator: Alex Pang

First published in Great Britain in 1999 by Macdonald Young Books, an imprint of Wayland Publishers Ltd
This edition updated and published in 2005 by Hodder Wayland, an imprint of Hodder Children's Books

Oxlade, Chris
Electricity and Magnetism. – (Science Files)
1.Electricity – Juvenile literature 2.Magnetism – Juvenile literature
I.Title
537

ISBN 075024710X

Printed in China by WKT Company Ltd

Hodder Children's Books
A division of Hodder Headline Limited
338 Euston Road, London NW1 3BH

Cover picture: Green electric light patterns.
Endpaper picture: Electricity supply cables held up by pylons.
Title page picture: Inside a television studio.

We are grateful to the following for permission to produce photographs: Digital Stock 9 top (Marty Snyderman), 16 bottom left (Bud Freund), 39 right; Digital Vision 18/19, 23, 29 bottom, 30 bottom, 34, 41, 42 top, 43 bottom; Getty images/Photodisc Red *cover* (Royalty-free/Steve Cole); Science Photo Library 8 (Doug Martin), 10 bottom (Peter Menzel), 12 bottom (David Parker), 13 (Kent Wood), 14 bottom (David Nunuk), 18 top and 18 bottom, 19 bottom (Phillippe Plailly/Eurelios), 20 (Charles D. Winters), 21 middle (Richard Megna), 26 bottom (Alex Bartel), 29 top (Francoise Sauze), 31 (John Mead), 33 top (Hank Morgan), 35 top (James Holmes), 37 bottom (Simon Fraser), 38 top (BSIP VEM), 38 bottom (Malcolm Fielding, The BOC Group plc), 41 top (BSIP S&I), 41 bottom (Will and Deni McIntyre), 42 middle (David Scharf); The Stock Market 5, 9 bottom (Jeff Zaruba), 25 bottom; Tony Stone 17 (Marc Dolphin), 25 top (Marc Pokempner), 36 (Mary Kate Denny).

Contents

Words in **bold** can be found in the glossary on page 44.

Introduction

Electricity and magnetism play a very important role in our modern lives. They work hundreds of machines, from torches to computers, and help us with hundreds of jobs, from sucking up dust with vacuum cleaners to flying airliners.

HISTORY FILE

MYSTERIOUS ATTRACTION

- About 2,500 years ago the ancient Greek philosopher Thales found that a lump of amber would attract dust and feathers if it was rubbed with a cloth. Today we call this effect **static electricity** (see pages 12-13).
- About 2,400 years ago in China, Emperor Huang-Ti used lumps of magnetic rock called lodestone as simple compasses. The lodestone was placed on a wooden float in a bowl of water. The lodestone always turned to point north-south.
- For the next two thousand years scientists did not get any closer to understanding electricity and magnetism. Their mysteries were finally understood in the eighteenth century.

WHAT ARE ELECTRICITY AND MAGNETISM?

Electricity is a form of energy. Energy comes in many other forms, too, such as light, heat and sound. Nothing could happen without energy to make it happen. Electricity is a very convenient form of energy because it is easy to change into other forms of energy. This is why we use electricity to move energy from place to place.

Welding machines need electricity to produce heat for joining metal.

Magnetism is a force. It attracts (pulls towards) or repels (pushes away) iron and some other substances. For example, a fridge magnet is attracted to the steel case of the fridge.

ELECTRICITY AND MAGNETISM IN NATURE

We normally see electricity and magnetism at work in machines, lights and so on, but they can also be seen in nature. For example, a flash of lightning is a giant spark of electricity, and a lump of rock called lodestone is a natural magnetic substance.

ELECTRICITY AND MAGNETISM WORK TOGETHER

Electricity and magnetism are very closely linked. In fact, electricity can cause magnetism and magnetism can cause electricity. For example, when electricity flows through a wire, it creates a magnetic field around the wire. And moving a wire near a magnet makes electricity flow in a wire.

Electricity and magnetism often work together. They produce a force called electromagnetism. This is one of the basic forces of the universe. The other basic forces are **gravity**, and two forces that are only found inside **atoms**.

Electromagnetism is everywhere today. Any device with an electric motor relies on electromagnetism; hospital scanners work with electromagnetism; and communications systems work with **electromagnetic waves,** such as radio waves and microwaves.

A shark can detect tiny pulses of electricity given out by its prey.

A TV studio control room relies on electricity and magnetism.

Electricity and Atoms

Electricity is created by atoms, so we need to know about atoms before we can understand electricity. Atoms are the particles that make up all substances. They are extremely tiny. There are millions and millions of atoms in a pin head.

Protons and neutrons make up the nucleus

Electrons move around the nucleus

An atom is made up of a nucleus and electrons.

Atoms themselves are made up of even more tiny particles called neutrons, protons and electrons. Neutrons and protons make up the **nucleus** of an atom. The nucleus is surrounded by a cloud of moving electrons.

Scientists use a Van der Graaf generator to build up huge electric charges.

Electric charge

Protons and electrons each carry a small amount of electricity, which is called an **electric charge**. But the charges are opposite to each other. A proton has a positive charge and an electron has a negative charge. Neutrons have no charge. Atoms normally have the same number of protons and electrons. The charges on them cancel each other out, meaning that the atoms have no overall charge.

Charging objects

Objects, like the atoms they are made of, are normally electrically neutral. But sometimes an object loses or gains some electrons. For example, if you rub a plastic ruler with a woollen cloth, electrons jump from some atoms in the ruler to atoms in the cloth. The ruler loses electrons, so is left with an overall positive charge. The cloth gains electrons, so is left with an overall negative charge.

TEST FILE

FUN WITH CHARGES

- Rub a party balloon against your clothing. The balloon will become negatively charged and the clothing positively charged. The balloon will cling to the clothing. A similar thing happens if you comb your hair again and again with a plastic comb.
- Rub a party balloon on wool. This will give the balloon a negative charge. Hold the balloon close to some small pieces of paper. The negative charge attracts positive charges to the edges of the paper, so lifts the pieces up.

ATTRACTION AND REPULSION

Electric charges push and pull on each other. A positive charge and a negative charge always attract (pull towards) each other.

Two negative charges or two positive charges repel (push away) each other. The force gets larger as the charges get closer to each other.

FACT FILE

ELECTRIC CHARGE

- The amount of electric charge is measured in coulombs (C). One coulomb of negative charge is made up of 6 million million million electrons!
- A capacitor is a device that stores electric charge. A capacitor can release its charge very quickly.
- Capacitors are used in many electronic circuits, including those in cameras, radios and computers.

Two balls with opposite charges attract each other.

Two balls with negative charges repel each other.

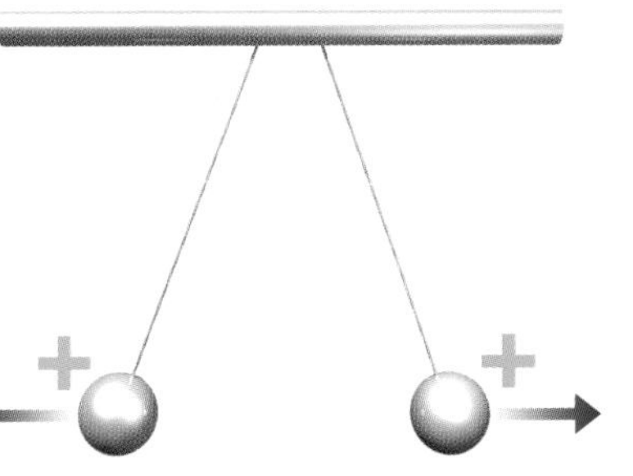

Two balls with positive charges repel each other.

Static Electricity

The electricity in an object that is electrically charged stays on the object's surface. It can jump from one object to another, but cannot flow through objects. It is called static electricity.

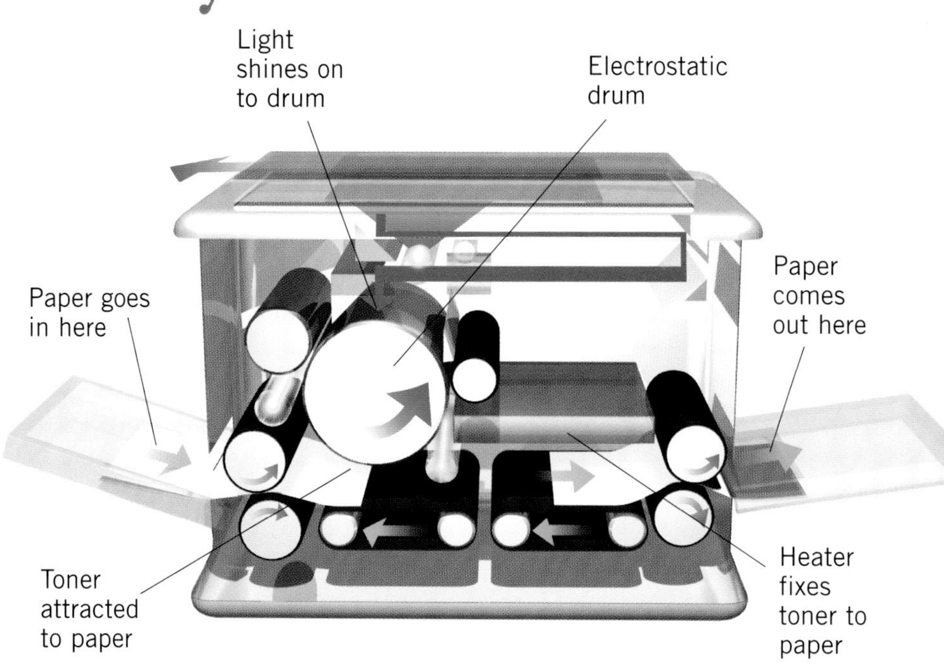

A photocopier works by using electric charges.

Using static electricity

We use the attractive and repulsive forces of static electricity to help tiny particles stick to surfaces. For example, a photocopier uses static electricity. Inside a photocopier is a drum. A system of mirrors and lenses bounces light off the document to be copied. This light then hits the drum, making a pattern of electric charge around it. The charge attracts tiny particles of black powder called toner, which have the opposite charge. This makes a black pattern of toner on the drum. The toner is then transferred from the drum on to paper and heated to make it stick.

The tiny particles of paint are all given the same electric charge as they come out of the nozzle. This makes them repel each other, which makes the paint spread evenly.

Losing charge

If the charge on an object becomes very large it can easily jump to an object close by that has no charge, or that has the opposite charge. This sudden rush of charge is called a **static discharge.** It often causes a spark.

Lightning

A bolt of lightning is a giant spark of static electricity. Inside a towering thundercloud, ice crystals bash into each other and become charged with electricity. Huge charges build up inside the cloud. Sheet lightning happens when a charge jumps from one part of a cloud to another. Forked lightning happens when a charge from a cloud moves through the air to the ground. Thunder is the noise made as the spark heats the air suddenly, making it expand.

TEST FILE

HOW FAR TO THE STORM?

Light travels a million times faster than sound. We see lightning almost straight away, but the sound of thunder takes longer to reach us. You can work out how far away a storm is by counting the seconds between lightning and the thunder. Divide the number of seconds by three to find the distance in kilometres.

FACT FILE

LIGHTNING

- Lightning heats the air to 30,000 °C. That's five times hotter than the surface of the Sun!
- A medium-sized lightning bolt contains enough electricity to supply a large town for a year.
- At any moment in time, there are about 2,000 thunder storms raging in different parts of the world.

A lightning bolt is made by static electricity discharging from a cloud.

Current Electricity

Current electricity is different to static electricity. Static charges stay on the outside of objects. But an electric current is made up of a stream of electrons flowing through a material.

Plastic case

Cells push electricity around circuit

Switch opens or closes circuit

Moving electrons

Imagine a long line of people, each with a tennis ball. Now imagine that each person passes their ball to the person behind, and gets a ball from the person in front. The balls move down the line, but the people stay still. Finally, imagine that each person is an atom, and each ball is an electron. When the electrons move from atom to atom, an electric current flows.

Bulb

The circuit in a torch contains cells, a switch and a bulb.

Very powerful currents flow along electricity supply cables.

Conductors and insulators

Electricity can flow easily through some materials. These materials are called **conductors.** Most metals are good conductors. Inside a conductor, some electrons are loosely connected to their atoms. They can easily move from atom to atom to make an electric current.

An **insulator** is a material that electricity cannot flow through. Plastics, paper, wood, glass and ceramics are all insulators. So are gases, such as the air. In an insulator, electrons are held tightly by their atoms, so they cannot move. We use insulating materials to stop electricity flowing where we don't want it to. For example, wires are covered with insulation plastic to stop the electricity flowing to other wires.

A few materials can work as both conductors and insulators. They are called **semiconductors**, and are used to make electronic devices such as **transistors** and **microchips**.

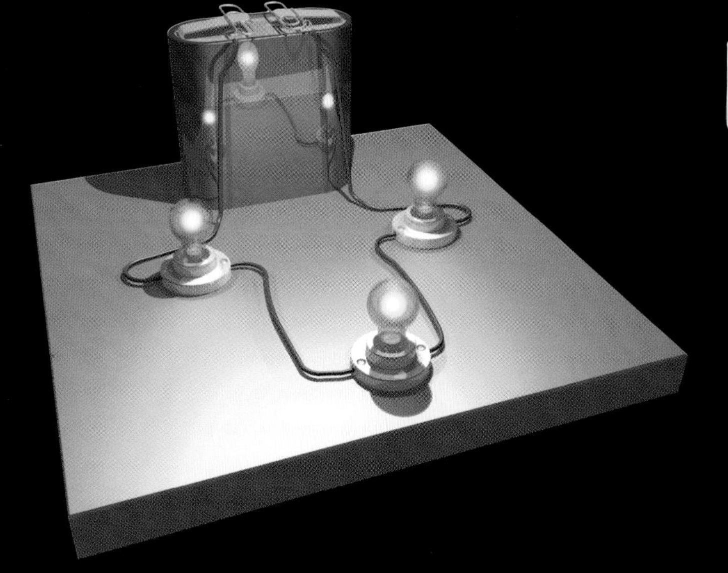

A series circuit. The current flows through the bulbs one after the other.

FACT FILE

CIRCUITS AND CURRENTS

Series circuit
In a series circuit the components are arranged in a chain in a single loop. The electric current passes through the components one after the other. The same amount of current passes through each one. If one component fails, the circuit is broken and none of the components work.

Parallel circuit
In a parallel circuit the components are arranged in separate loops. The current divides up to pass through all the components at the same time. If one component fails, the circuit is not broken and continues to work.

A current can also flow around a circuit in two different ways.

Direct current (DC)
A **direct current** flows around a circuit in the same direction all the time. Direct current flows in circuits in torches and other small devices.

Alternating current (AC)
An **alternating current** flows one way round a circuit and then the other. It changes direction many times a second. Alternating current is used in mains circuits in your home.

A parallel circuit. The current flows through all the bulbs at the same time.

MAKING A CIRCUIT

An electric current can only flow in a continuous loop made of conducting material such as copper wire. This loop is called a circuit. Electrons don't flow round a circuit on their own. They need a push, which is called an **electromotive force (emf).** In a torch, emf comes from a battery.

COMPONENTS

Bulbs, switches, motors and other devices in electric circuits are called components. Each component does a particular job in the circuit. For example, a bulb gives out light when current flows through it, and a switch makes or breaks a circuit, allowing current to flow or stopping it flow.

Cells and Batteries

Cells and batteries, like the ones we put in torches and radios, are like stores of electricity. They make an electromotive force (see page 15) that pushes current around circuits. The scientific name for some of the types of batteries we use is 'cells'. A battery is actually made up of two or more cells joined together.

A simple cell is made up of a chemical called an **electrolyte** and two **electrodes**, which are in the electrolyte. The two electrodes are made of different materials, which are often metals. They are connected to the two terminals on the outside of the cell, through which current leaves and returns to the cell.

When a cell is connected to a circuit, chemical reactions happen between the electrolyte and the electrodes. This makes electrons flow out of the cell's negative terminal, round the circuit, and back into the positive terminal.

The rechargeable battery in a mobile phone supplies electricity for several hours.

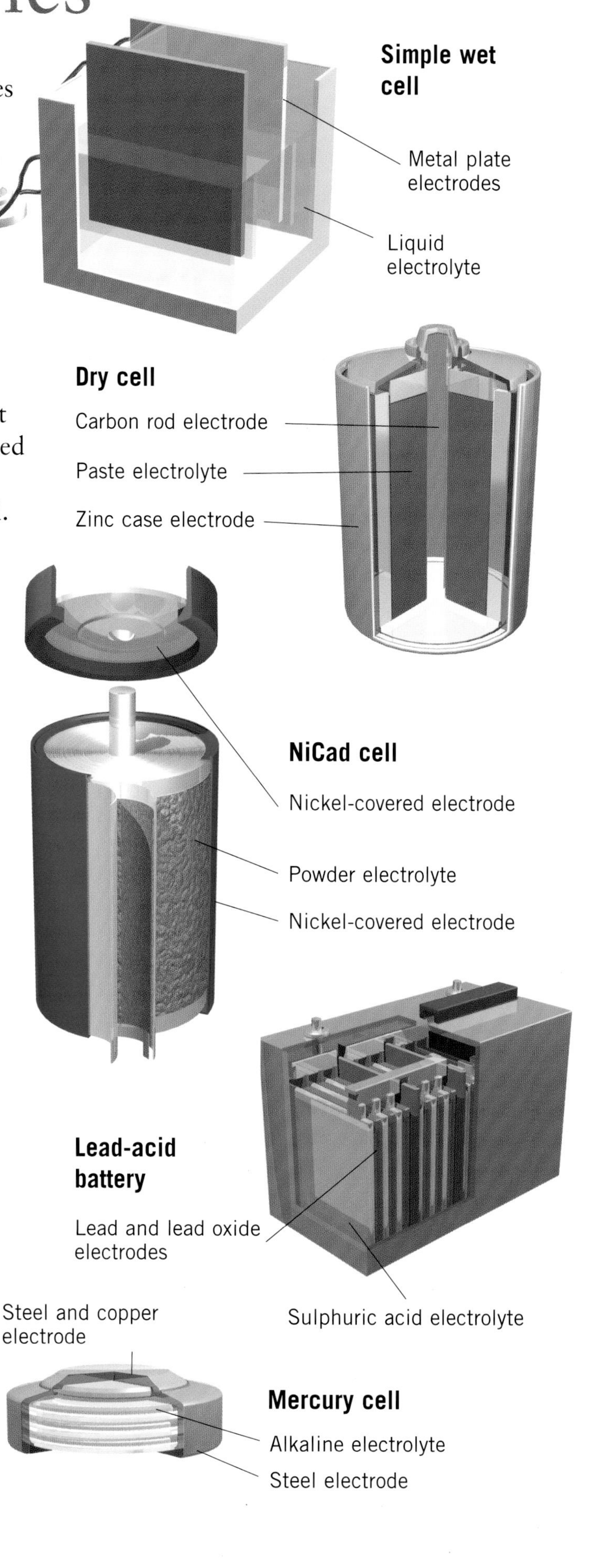

Portable devices such as personal CD players are powered by dry cells. A set of cells runs the device for a few hours.

PRIMARY AND SECONDARY CELLS

Cells are either primary cells or secondary cells. In a primary cell, the chemicals in the electrolyte and the electrodes are gradually used up as the cell gives out current. Eventually the chemicals run out and the battery dies. A secondary cell can be recharged when it stops giving out current. The reactions that happen inside are reversed when current is put into the battery, so the electrodes and electrolyte reform and the cell can be used again.

WET AND DRY CELLS

A wet cell is a cell in which the electrolyte is a liquid. We have to be careful with wet cells to stop the electrolyte escaping. Dry cells are much more convenient to use. We use them in all sorts of portable devices, such as torches and MP3 players. The electrolyte in a dry cell is either powder or paste.

A common type of dry cell is the zinc-carbon cell. One electrode is a rod of carbon in the centre of the cell, and the other is the battery's zinc casing. The electrolyte is ammonium chloride paste. An alkaline cell lasts longer than a zinc-carbon cell. It uses potassium hydroxide, which is a type of chemical called an alkali, as its electrolyte. A NiCad cell is a rechargeable cell that contains nickel and cadmium. It is used in mobile phones, cordless tools and other devices.

LEAD-ACID BATTERIES

A car battery is a lead-acid battery. Its electricity is used to start the car's engine, and it is recharged when the engine is running. A lead-acid battery is made up of several rechargeable wet cells. Each cell is made up of lead and lead oxide sheets in a sulphuric acid electrolyte.

HISTORY FILE

THE FIRST BATTERY

The first battery was invented in 1800 by the Italian physicist Alessandro Volta (1745-1827). It was made of a pile of zinc and copper discs separated by paper discs soaked in salt water. Simple batteries like this were the first source of electric current for scientific experiments.

Measuring Electricity

We need to be able to measure electricity so that we can see how much electricity is flowing in a circuit, and also to choose cells, bulbs and other components. Here you can see the main electrical measurements used by scientists. They are named after scientists who studied electricity and other forms of energy in the past.

Alessandro Volta

Volts (V)

Units called volts (V) are used to measure electromotive force (emf). For example, a normal torch battery produces an emf of 1.5 volts. Domestic electricity supplies have an emf of 110 volts or 240 volts. Electricity distribution cables carried on pylons have an emf of up to 400,000 volts, or 400 kilovolts (kV). The volt is named after the Italian physicist Alessandro Volta (1745-1827).

Georg Ohm

Amperes (A)

The size of a current is measured in amperes or amps (A). A current of 1 amp means that 1 coulomb of charge (see page 11) is passing a point in a circuit every second. Only a fraction of an amp flows around the circuit in a torch. An electric kettle uses about 4 amps. The amp is named after the French scientist André-Marie Ampère (1775-1836).

Ohms (Ω)

The ohm (Ω) is used to measure electrical **resistance.** Resistance is how much an object tries to slow down the current flowing through it. A copper wire has a resistance of just a few ohms. The filament inside a light bulb has a resistance of a few hundred ohms. The ohm is named after the German scientist Georg Ohm (1789-1854).

Engineers need to measure electricity to repair circuits.

FACT FILE

USEFUL EQUATIONS

Two simple equations show how voltage, current, resistance and power are linked together in a circuit and in components in circuits.

- The difference in volts between one part of a circuit and another is equal to current (in amps) times resistance (in ohms), or $V = A \times \Omega$.

- The power in watts used by a circuit or a component is equal to the emf (in volts) times the current (in amps), or $W = V \times A$.

WATTS (W)

The watt (W) is the unit of **power**, which is the amount of energy used in a second. In an electrical device, power measures how much electrical energy it uses every second to work. A normal light bulb uses 60 or 100 watts. A microwave oven uses about 800 watts on full power. A kilowatt (kW) is equal to 1,000 watts. The watt is named after the Scottish engineer James Watt (1736-1819).

Scientists experimenting with *superconductors*, which have zero resistance.

Magnetism

Magnetism is a force that acts between objects called magnets. It can attract objects to each other, or push them away from each other. Magnets are made in many different shapes. Most common are bar magnets, horseshoe magnets, disc magnets and ring magnets. Most are made from iron or contain iron. The area around a magnet where its force can be felt is called a **magnetic field.**

TEST FILE

MAGNETIC EXPERIMENTS

• See a magnetic field

Put a piece of paper over a magnet. Sprinkle iron filings over the paper. The filings line up with the lines of force in the magnet's magnetic field.

• Find magnetic materials

Hold a magnet next to objects made from various different materials. If you can feel the magnet being attracted to the object, the object is made of magnetic material. Don't hold the magnet next to a computer because you could damage the hard drive.

• Make a magnet

Bend a steel paper clip to make it straight. Stroke it from end to end with one end of a bar magnet. Always stroke it in the same direction, and lift it clear of the clip to return to the other end each time. This will make the paper clip into a magnet.

A magnet attracting steel scraps.

Magnetic materials

Materials that can become magnets or that are attracted to magnets are called magnetic materials. There are only a few magnetic materials. The most common magnetic material is the metal iron. Steel is also magnetic because it is made mostly of iron. So is the rock lodestone, which contains iron oxide. The metals cobalt and nickel are also magnetic, but most metals are not.

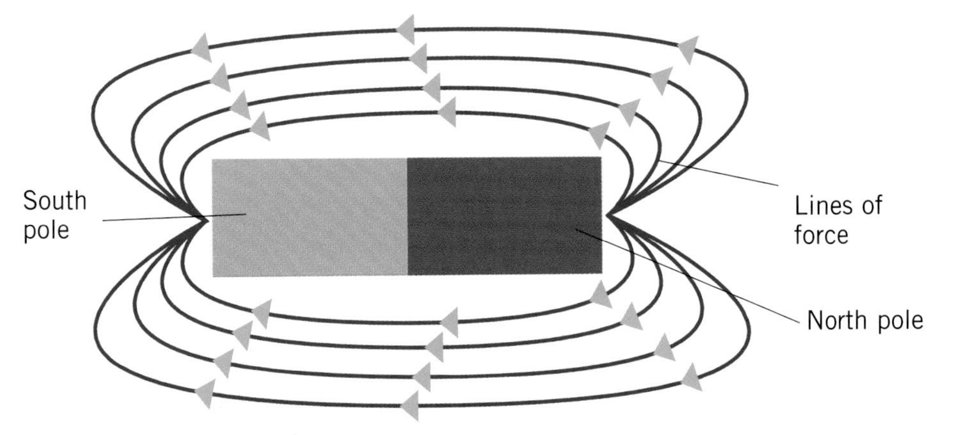

The lines of magnetic force start and finish at the magnet's poles.

Magnetic poles

A magnet always has two points where its magnetic force is strongest. These are the magnet's poles. Lines of magnetic force come out from the poles. The Earth has a magnetic field similar to a bar magnet, with its poles near the geographic poles. If a magnet is allowed to spin round, its poles always point towards the Earth's poles. The pole that points north is called the north pole, and the pole that points south is called the south pole.

Opposite poles on magnets (a north and a south) always attract each other. Like poles (two north or two south) always repel each other.

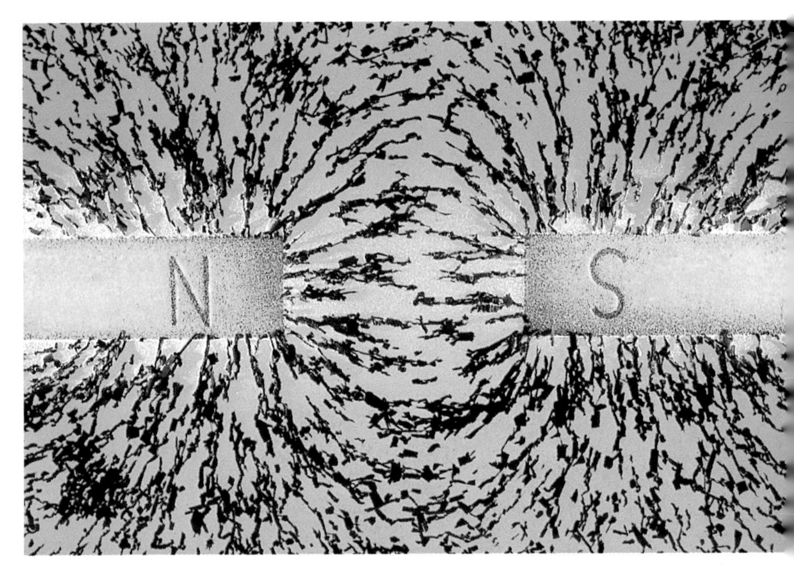

Sprinkled iron filings show up the lines of force between two magnetic poles.

What makes things magnetic?

Magnetism happens because of the atoms in a material. An atom is made up of a nucleus and electrons that move around it. Magnetism is caused by the movement of the electrons.

Groups of atoms in a material, all with the same magnetism, are called **domains**. Each domain is like a tiny magnet. In a magnetic material that is not a magnet, the domains are jumbled up. In a magnet, they all point in the same direction.

Stroking an iron bar with a magnet turns it into a magnet. Hammering it destroys the magnet.

Hammer blow jumbles domains again

Domains jumbled up

Domains lined up

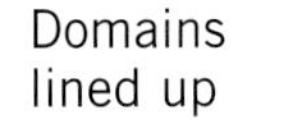

Using Magnets

We use magnets for dozens of different jobs at home and in industry. The simplest application of magnets is to hold things temporarily. For example, we use fridge magnets to fix notes to metal surfaces, and a screwdriver has a magnetic holder that holds screwdriver bits and screws in place.

A burglar alarm has magnetic sensors on outside doors. Each sensor has a magnet on the door and a switch on the door frame. The magnet keeps the switch closed when the door is shut. If the door opens, the switch opens, triggering the alarm.

Finding the way

A magnetic compass is a device for finding the way. Inside is a needle that is a bar magnet. The needle always swings round to point to the Earth's magnetic north pole. Walkers and sailors can find the direction they want to go in by looking at the needle.

Maglev (magnetic levitation) trains have no wheels. They are held above the track by magnetic repulsion.

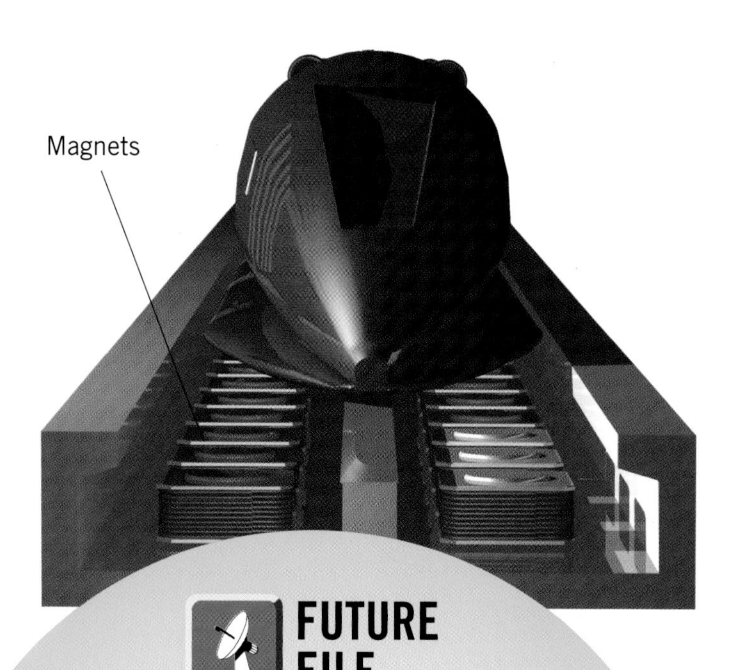

FUTURE FILE

FLOATING TRAINS

For many years, railway engineers have been experimenting with magnetic levitation (maglev for short) trains. Powerful magnets keep the train floating above the track, and other magnets pull it along. There is no **friction** with the track, which allows the train to travel faster than trains with wheels.

Magnetic attraction pulls the train along the track.

This iron ore was found using sensitive magnetic instruments.

Finding minerals

The force lines of the Earth's magnetic field are not always perfectly straight. In some places they are bent or distorted. These bends often show that there are rocks containing iron under the surface. Scientists use instruments to measure the field's direction, and so find valuable iron ore.

Smooth spinning wth magnets

A bearing is a device that allows a wheel or axle to spin smoothly. Most bearings contain metal balls or rollers covered with oil that reduce friction. Magnetic bearings are better than these ball bearings. They use magnetic repulsion to keep the parts of the bearing apart. This cuts out friction completely.

Magnetic information

Video tapes, cassette tapes and computer discs all use magnetism to store information, such as video pictures, sounds and computer data. The tapes are made of plastic, coated with magnetic material. During recording, a magnet makes a magnetic pattern on the tape. During playback, the pattern is read from the tape by an electronic circuit.

FACT FILE

MEASURING FIELDS

The strength of a magnetic field is often measured in units called teslas (T). The higher the strength, the higher the number of teslas. Here are some measurements for different magnetic fields.

- Around a fridge magnet: 0.02 teslas.
- Around a hand-held bar magnet or horseshoe magnet: 0.1 teslas.
- The Earth's magnetic field: 0.00005 teslas.
- Around a neutron star (a collapsed dead star): 100 million teslas.

Electromagnetism

When an electric current flows along a wire it makes a magnetic field around the wire. This effect is called the **electromagnetic effect.** It is caused by the electrons of the electric current moving along the wire.

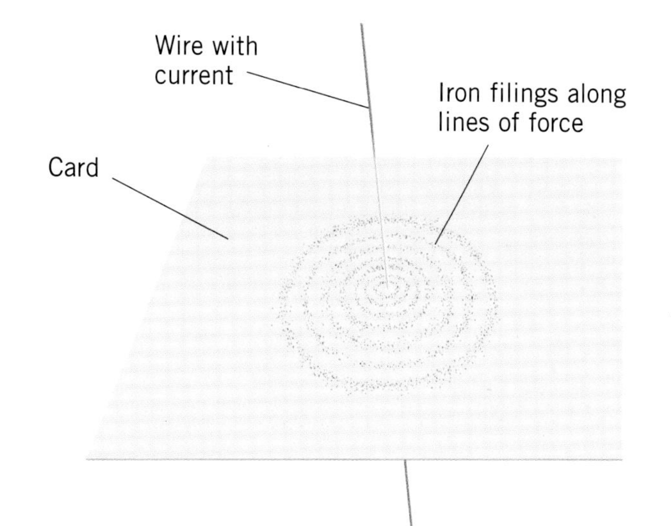

Sprinkling iron filings on paper around a current-carrying wire shows up the rings of the magnetic field.

TEST FILE

MAKE AN ELECTROMAGNET

- Wrap a length of insulated copper wire tightly around an iron or steel nail or bolt. Leave about 20 cm of spare wire at each end. Use sticky tape to hold the coils in place.
- Wrap the bare ends of the wire around two paper clips, and slip the clips on to the terminals of a 4.5 volt battery.
- Try picking up small steel pins or staples with your electromagnet.
- Unclip one wire and the magnet loses its strength.
- Make a switch for the electromagnet with drawing pins, a paper clip and a wooden block.

ELECTROMAGNETS

The magnetic field around a single wire is very weak. It can be made much stronger by winding the wire round and round to make a coil called a solenoid. Wrapping the coil round an iron rod makes the field stronger still. The coil and rod is called an electromagnet. Its magnetism can be turned on and off by turning the current in the coil on and off. It has poles at each end, too, just like a permanent magnet.

Paper clip switch
Steel nail
Coil of wire

A steel nail with a wire coil around it becomes an electromagnet.

Making electricity

If a wire with current flowing through it is put in a magnetic field, the wire moves to one side because it is attracted or repelled by the magnet. This effect works in reverse, too. For example, if a wire moves through a magnetic field, a current flows through the wire. This is called **electromagnetic induction.**

An electric guitar works by electromagnetic induction.

HISTORY FILE

OERSTED'S DISCOVERY

Hans Christian Oersted (1777-1851) was a professor at Copenhagen University in Denmark. While he was giving a lecture on electricity in 1820, he noticed that the needle in a compass near a wire was twitching as he turned the electricity in the wire on and off. He realised that the electric current was creating a magnetic field.

Play it again

The strings of an electric guitar make only a very quiet sound on their own when they are plucked. But their vibrations are detected by electromagnetic induction, which creates currents that are amplified and played through a loudspeaker. Under each metal string is a device called a pick-up, made up of magnets with coils of wire around them. When the string vibrates, it changes the magnetic field of the magnet, which makes a current in the wires.

An airport metal detector senses the change in a magnetic field caused by metal objects.

Using Electromagnets

Electromagnets are very useful devices that are used in machines from cranes to loudspeakers. They are more useful than ordinary, permanent magnets. An electromagnet can be turned on and off, and its strength can be changed by changing the size of the current flowing through it.

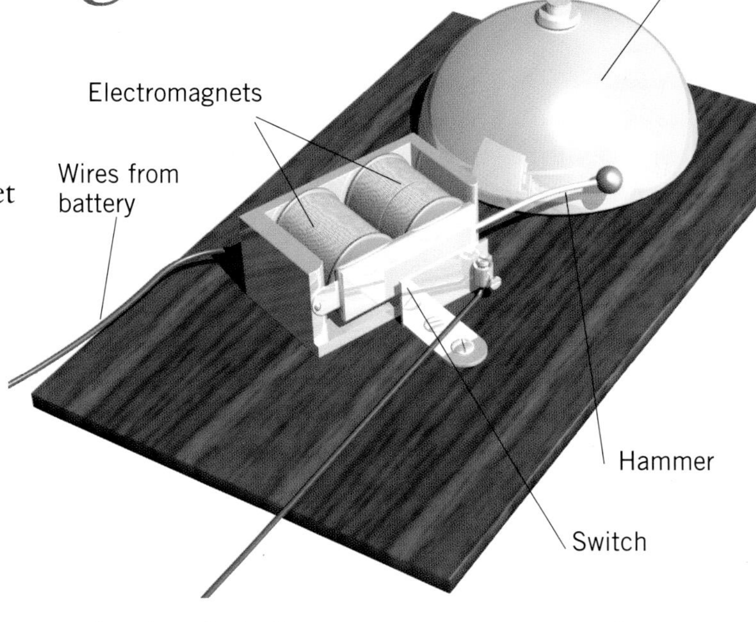

An electric bell uses electromagnets to ring.

Lifting with electromagnets

Powerful electromagnets create very big magnetic forces. They are used in scrapyards and recycling centres to pick iron and steel from heaps of mixed scrap.

A strong electromagnet separating iron and steel from other metal scrap.

Electromagnetic bells

Some doorbells contain an electromagnet. When the doorbell button is pressed, an electric current flows through the electromagnet, turning it on. The electromagnet pulls the hammer against the bell, making a clang. At the same time it opens a switch in the circuit. This turns off the electromagnet. Then the hammer springs back and the switch closes again. The process is repeated again and again, making the bell ring.

Electromagnetic locks

Another device on a door that uses an electromagnet is a remote-control door lock. The sliding door latch is normally pushed into the door frame by a spring, keeping the door locked. When the switch inside the building is pressed, electricity flows through a solenoid (see page 24) in the lock that surrounds the latch. The latch is pulled into the solenoid and out of the frame, releasing the door.

Sound to electricity

Sound is made up of vibrations that spread through the air. The job of a microphone is to turn sound into an **electrical signal** that represents the sound. Inside the microphone is a coil of wire that is inside the magnetic field of a magnet. The coil is attached to a thin disc called a diaphragm. When the vibrations of sound hit the diaphragm, it vibrates too. This makes the coil move inside the magnetic field. Electromagnetic induction causes a changing current to flow in the coil. This changing current mirrors the vibrations in the sound waves.

FUTURE FILE

INTO SPACE BY MAGNETS

Launching satellites into orbit around the Earth is very expensive. The rockets that launch them cannot be used again, and huge amounts of fuel are used. One idea for launching satellites is to use a giant electromagnetic gun, full of ring-shaped electromagnets. This would use powerful electromagnets to accelerate the satellite to such a high speed that it would be hurled into space.

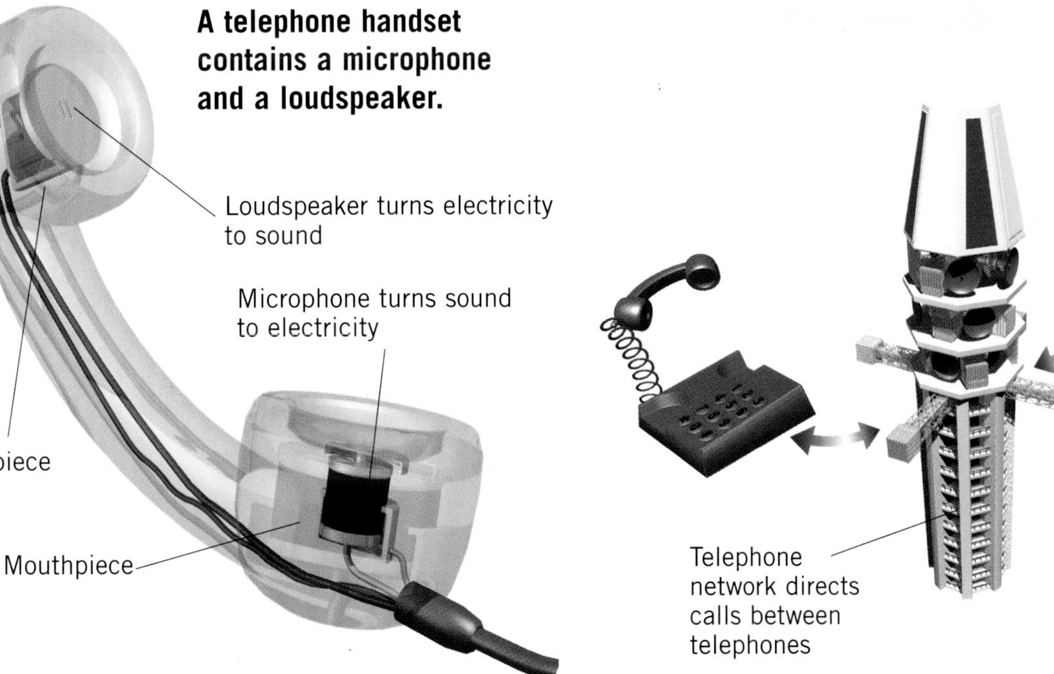

A telephone handset contains a microphone and a loudspeaker.

Electricity to sound

A loudspeaker does the opposite job to a microphone. It turns an electrical signal (a changing current that represents a sound) into the sound itself. Inside a loudspeaker is a coil of wire that is inside the magnetic field of a magnet. The coil is attached to a cardboard cone at the front of the speaker. When an electrical signal flows through the coil, the coil becomes an electromagnet and moves in and out in the magnetic field. This makes the cone vibrate in and out, which makes the sound. A telephone handset contains both a microphone and a loudspeaker.

Electric Motors

One of the most important applications of electromagnetism is the electric motor. A motor turns the energy in an electric current into spinning movement. Electric motors are used to drive thousands of different machines.

The simplest electric motor is made up of a coil of wire mounted on a shaft, with permanent magnets around it that create a magnetic field across the coil. Electricity gets to the coil through contacts called brushes, which keep contact as the coil rotates.

When a current flows through the coil, the coil becomes a magnet. The permanent magnets force one side of the coil to move upwards and the other side to move downwards, until the coil has completed a half turn. Now the direction of the current is reversed. This reverses the magnetism of the coil, and the coil makes another half turn, completing the turn.

Direct current (DC) motors

The electric motors in small devices such as toys and CD players are powered by direct current (DC) from cells and batteries. Direct current always flows in the same direction. A device called a commutator inside the motor is needed to change the direction of the current after every half turn.

Cell or battery
Commutator
Brush contact
Motor shaft
Permanent magnet
Coil of wire

A simple DC motor. The commutator reverses the current as the coil spins.

HISTORY FILE

MILESTONES IN ELECTRIC MOTORS

1821: In an experiment, English scientist Michael Faraday (1791-1867) made a wire with a current inside spin around a magnet.

1831: American scientist and engineer Joseph Henry (1797-1878) made the first practical electric motor.

1870s: Factories started to use electric motors to work their machinery.

1888: Croatian-American engineer Nikola Tesla (1856-1943) designed the first induction motor.

FACT FILE

MOTOR EFFICIENCY

Electric motors are very good at changing the energy in electricity into movement. They hardly waste any energy. We say that they are very efficient. Other engines are much less efficient.

- A steam engine changes only about 20 per cent of the energy in its fuel into movement.
- For a petrol engine the amount is about 40 per cent.
- For an electric motor, the amount is more than 90 per cent.

High-speed trains have efficient electric motors.

Alternating current (AC) motors

The electric motors in machines such as washing machines, vacuum cleaners and hairdryers use alternating current because the mains supply is alternating current. An alternating current changes direction many times a second. This means that an AC motor does not need a commutator as a DC motor does. These motors also have many coils instead of just one, which means they turn very smoothly instead of in jerky movements.

An induction motor is a type of AC motor with no brushes. Instead of magnets around the coil, it has more coils. Changing currents in these coils create currents in the inner coil. This makes the inner coil spin.

Most power tools, such as this circular saw, have AC motors.

Generating Electricity

An electric motor turns the energy in electricity into movement. A device called a **generator** does the opposite job. It turns movement into electricity. Generators are used to produce electricity in power stations, cars, wind-up radios and similar devices. The inside of a simple generator looks very much like the inside of an electric motor. There is a coil of wire on a spindle and permanent magnets fixed around it. When the spindle is turned the coil moves through the magnetic field of the magnets. Electromagnetic induction makes current flow in the coil. The current comes out of the coil through brushes.

A simple DC electrical generator used to light a bulb. It is very similar to a DC motor (see page 28).

AC and DC generators

As the coil in a generator turns, the wires on each side pass up through the magnetic field and then, half a turn later, down through the magnetic field. This makes current flow one way in through the coil and then the other way out through the coil. So the current that comes out of the generator flows one way then the other. It is alternating current (AC).

Hundreds of wind turbines together generate as much electricity as a small coal-fired or oil-fired power station.

A hydroelectric power station. The turbines are at the base of the dam.

FACT FILE

POWER STATION GIANTS

- An average coal-fired power station produces about 1,000 million watts (or 1 megawatt (MW)) of power.
- The Itaipu Power Complex of hydroelectric power stations between Brazil and Paraguay produces 13,320 MW. It is the world's most powerful power station.
- In 2009, the Three Gorges hydroelectric scheme in China will be able to produce 18,200 MW.
- The largest wind turbines produce about 4 MW each. There are plans to build 1,000 MW wind farms.
- The solar power plant at Harper Lake in the Mojave Desert, USA, produces 160 MW.

To make a generator produce direct current (DC), it needs a commutator like the one in a DC motor. This reverses the connections from the brushes to the coil every half turn of the shaft so that the current from the generator always flows the same way.

Generators in power stations

Electricity generating stations (or power stations) have huge generators that produce electricity for towns and cities. In different types of power stations, the generator coils are turned in different ways.

In a coal-fired or oil-fired power station, the burning fuel heats water that boils to make steam. The steam is fed to a device called a turbine, which contains many fan-like blades. The steam makes the blades spin. The turbine's shaft is connected to a generator. At a nuclear power station, the water is boiled by heat from a **nuclear reactor.**

At a **hydroelectric power** station, fast-flowing water from behind a dam makes a turbine spin. The turbine makes a generator spin. At a tidal power station, the flowing water comes from the tide flowing in and out. A wind turbine has a rotor like a propeller. The wind makes it turn and it turns a generator.

Electricity Distribution

The electricity we use at home, at school, in offices and factories, and for lighting streets is produced at electricity generating stations. It gets to us through an electricity distribution network.

Electricity travels from generating stations to towns and cities along thick cables that hang on pylons. The alternating current in the cables has very low current, but very high voltage. At high current and low voltage, it would make the cables hot, and energy would be lost. At a power station, the generators produce electricity at about 22,000 volts. Devices called **transformers** (see below) change the voltage to 400,000 volts for its journey along the cables.

Engineers monitor the flow of electricity through the network.

Transformers

A transformer is a device that changes the voltage and current of electricity. A simple transformer has two coils of wire wrapped around an iron core. When an alternating current flows in one coil (called the primary coil), it makes an alternating current flow in the other coil (the secondary coil), too. A transformer works by electromagnetic induction. The current in the primary coil turns it into an electromagnet, which creates the current in the secondary coil.

The inside of a high-voltage cable.

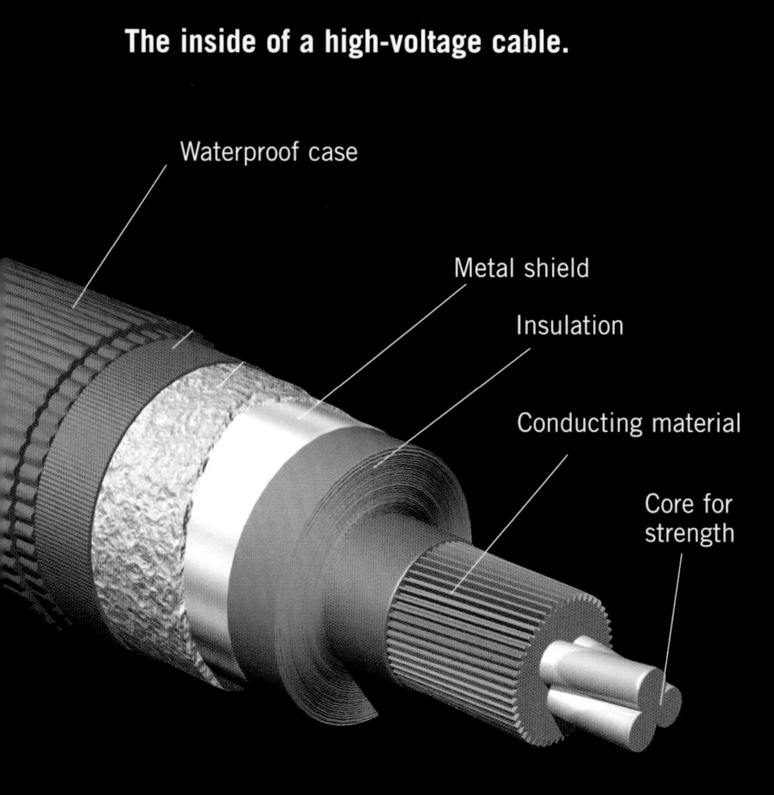

FACT FILE

HIGH VOLTAGE

Different electricity users need electricity at different voltages to operate different sorts of machinery. Here are some examples:

- Heavy industrial factories, such as steel works: 33,000 volts.
- Light industrial factories, such as electronics factories: 11,000 volts.
- Houses and homes: 110 volts or 240 volts.
- Small devices such as computers and CD players: 12 volts, 9 volts or less. These devices have their own step-down transformers.

Stepping up

The transformer at a power station is called a step-up transformer because it makes the voltage higher. In a step-up transformer, the secondary coil has more turns of wire than the primary coil. This makes the voltage higher, but the current smaller. For example, if there are three times as many coils in the secondary coils than the primary coils, the voltage becomes three times as high. But the amount of energy in the electricity stays the same.

Stepping down

Electricity from distribution cables has voltages that are too high to use in factories, offices, homes and schools. These voltages are extremely dangerous, which is why cables are held up on tall pylons. So the voltage must be reduced again. This is done by step-down transformers at electricity substations. A step-down transformer has fewer coils in its secondary than its primary coil. The voltage is stepped down about ten times for factories, and about two thousand times for homes and offices.

The electricity grid

Electricity generating stations, cables and substations are all linked together to make a network called a grid. Electricity can take many routes through the grid. Then if one part of the grid goes wrong, the electricity can be sent to people along another route.

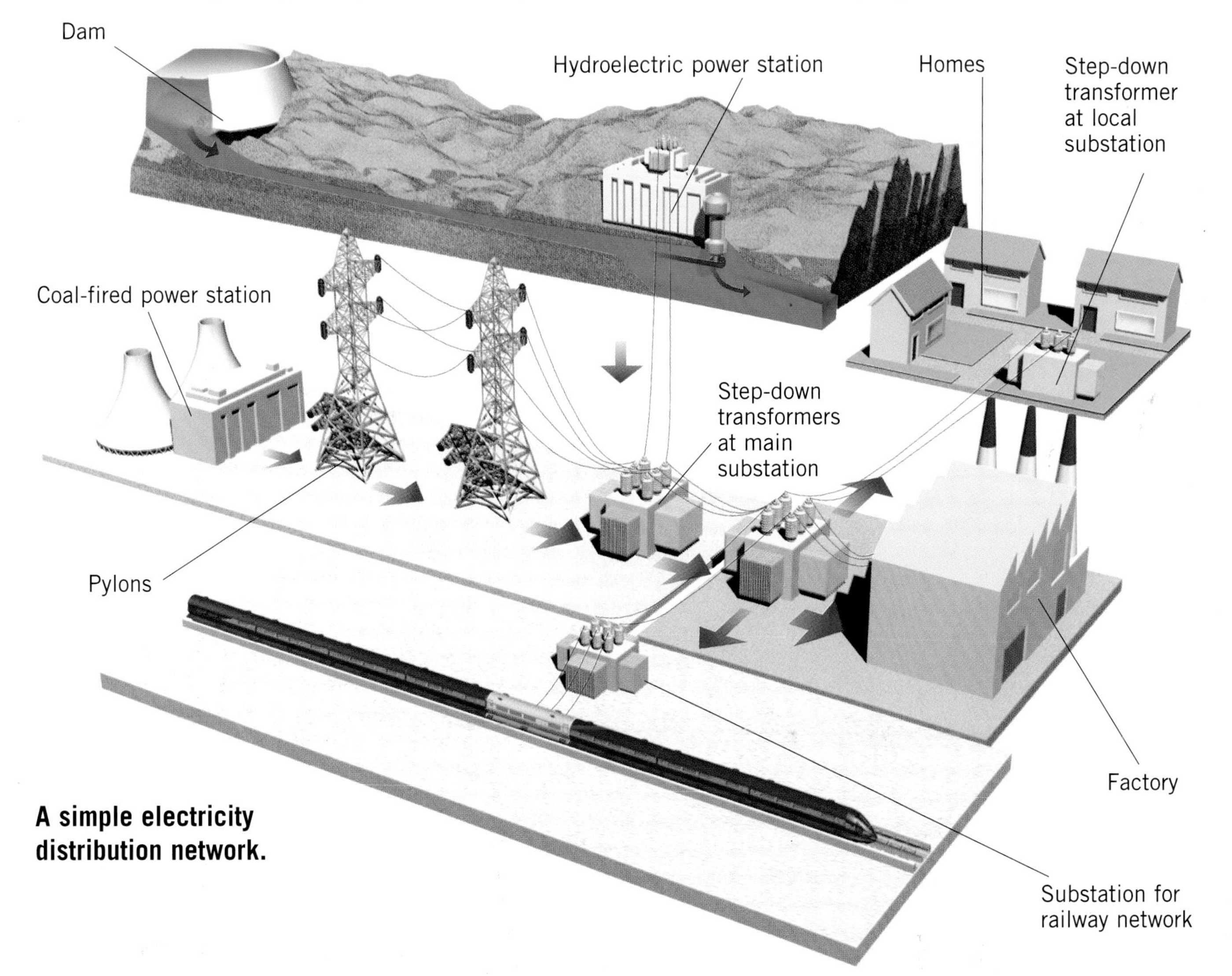

A simple electricity distribution network.

Electricity in Industry

Huge amounts of electricity are needed in a steel works.

Industries use up much more electricity than the electricity we use in our homes, schools and offices put together. So what do they use it for? The simplest industrial use of electricity is lighting. But lighting is usually the smallest part of a factory's electricity bill.

The next biggest use is for moving things around the factory. Machines such as conveyor belts, roller tracks, pulleys and pumps are all powered by electric motors. Cutting machines such as drills, lathes and grinders all have electric motors too.

Factories normally spend most of their electricity bill on heating the inside of the factory to keep the workers warm. Electricity is also needed to operate ovens, furnaces, moulds and kilns. The ovens in a large bakery use as much electricity as thousands of homes.

Many industrial processes also use electricity. One of the most important is **electrolysis**, which is used to split up chemicals.

Electrolysis

When some substances dissolve in liquid, they break up into particles called **ions**. An ion is an atom that has gained some extra electrons or lost some of its electrons. This means it has a negative or positive charge.

For electrolysis, two electrodes, which are normally made of metal, are put into the liquid containing the ions. One is a negative electrode, or cathode, and the other is a positive electrode, or anode. An electric current is passed between the electrodes. The negative ions move towards the anode and the positive ions move towards the cathode. There they lose or gain electrons to turn back to uncharged atoms. For example, during electrolysis of sodium chloride solution, positive sodium ions move to the cathode and become atoms of sodium metal.

TEST FILE

COPPER ON SPOONS

Warning: the copper sulphate used in this project is poisonous. Ask an adult for help and never drink the solution. Always wash your hands afterwards. You can get copper sulphate from a chemistry kit or from a pharmacy.

- Dissolve a tablespoon of copper sulphate crystals in a beaker of water.
- Connect an old metal spoon to one terminal of a 4.5 volt or 9 volt battery.
- Connect a piece of copper pipe to the other terminal.
- Hold the spoon and pipe in the copper sulphate solution.
- The spoon will get a shiny coating of copper.

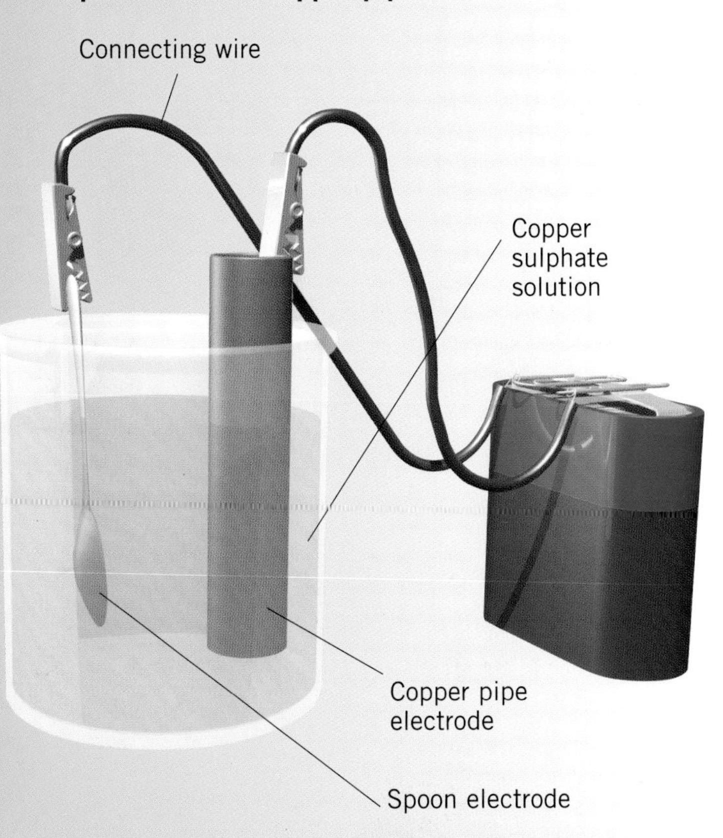

The copper on the spoon comes from the copper sulphate and the copper pipe.

These are electrolysis tanks that produce chlorine from salt water.

SMELTING ALUMINIUM

Smelting is the process of extracting metals from the chemicals they are found in, which are called **ores**. Aluminium is smelted by electrolysis from its ore, bauxite, which contains aluminium oxide. The aluminium oxide is heated to 1,000 °C to make it liquid containing aluminium ions. Electrolysis separates the aluminium ions and turns them into aluminium metal.

ELECTROPLATING

Electrolysis is also used to cover metal objects with a thin layer of another metal. This is called **electroplating**. The object is used as one of the electrodes in the electrolysis of a liquid that contains ions of the plating metal. The ions are attracted to the object and turn to atoms that coat it. Electroplating is used for silver plating ornaments and for coating objects with zinc to stop them rusting.

Electricity at Home

Our modern homes rely on electricity. We use electricity for lighting, for heating, for cooking, and to work dozens of different electric devices for washing, cleaning, entertainment and telling the time. It is only when our electricity is cut off for a while that we realise how much we depend on it!

WARNING! NEVER PLAY WITH THE MAINS ELECTRICITY SUPPLY. IT CAN EASILY KILL YOU.

BRINGING ELECTRICITY TO OUR HOMES

Mains electricity gets to our homes along thick cables from the local electricity substation. The cables run underground or in the air, held up by telegraph poles. When a cable enters a home, it goes to an electricity meter which measures how much electricity we use. Then it goes to a box called a consumer unit. Here it connects to different circuits that take electricity around the home. There are lighting circuits, circuits for wall sockets, a cooker circuit, and other appliances.

Each circuit has a device called a **fuse** or a circuit-breaker in the consumer unit. If something goes wrong with the circuit or a device plugged into the circuit, the current

Electricity heats water and works motors in a dishwasher.

FACT FILE

HOW MUCH ELECTRICITY?

This list shows how long different appliances would work on 3.6 million joules of electrical energy (enough to run a light bulb for 10 hours).

Instant shower	8 minutes
Electric double-oven	15 minutes
Electric kettle	30 minutes
Tumble-drier	40 minutes
Hand-held hairdryer	40 minutes
Microwave oven	45 minutes
Turbo-type vacuum cleaner	50 minutes
Hot-water immersion heater	1 hour
Ordinary vacuum cleaner	1 hour 30 minutes
Large television	4 hours
100-W light Bulb	10 hours
Small, portable television	12 hours
Small domestic music system	12 hours
Fluorescent low-energy light bulb	30 hours
Electric toothbrush	40-50 hours

could become dangerously large, and could hurt somebody. The fuse or circuit-breaker switches the electricity off instantly if there is a fault.

LIVE, NEUTRAL AND EARTH

Each circuit in a house has three wires inside. They are called the live, neutral and earth wires. The live wire carries electricity from the consumer circuit to the sockets, lights and appliances. The neutral wire completes the circuit. It carries electricity away again, back to the consumer unit. The earth wire is a safety wire. If something goes wrong with a circuit or an appliance, the earth wire carries the electricity safely into the ground.

FUTURE FILE

HOMES OF TOMORROW

In the future homes will be made more environmentally friendly by automatically saving electricity. An electronic system will monitor the weather, the temperature in the house, and which rooms are being used. It will adjust the lighting, heating and air conditioning so that as little electricity as possible is used.

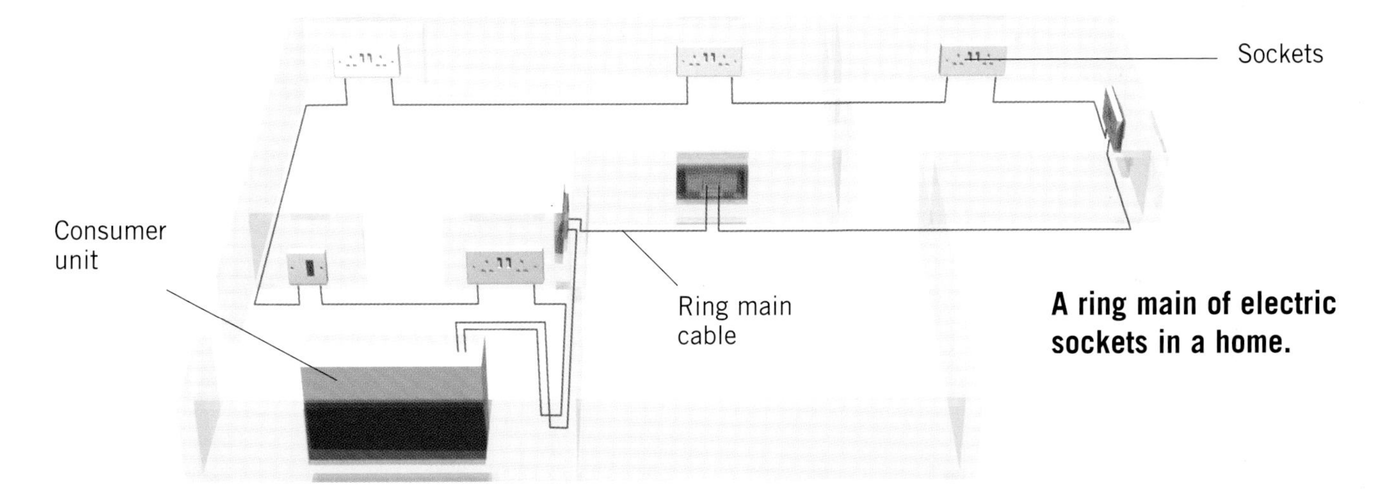

A ring main of electric sockets in a home.

RINGS OF CABLES

The circuits that carry electricity to sockets and lights are called ring mains. Their cables leave the consumer unit, go to the first socket on the ring, then to the next socket, and so on, and finally back to the consumer unit.

Demand for electricity is greatest in the early evening in winter when people need electricity for cooking, heating and lighting.

Electricity and Magnetism in Medicine

Our bodies are full of tiny electric currents. They are called nerve signals. They travel along the nerves that connect your brain to all the other parts of your body. Nerve signals carry information to your brain about how your body is working, and carry instructions from your brain that control your body. Even our thoughts are made up of tiny electric currents in our brains!

Sensors on the skin

Nerve signals can be detected by electronic sensors stuck on the skin near the nerves. The tiny signals are amplified and shown up on a screen or on paper. The pattern of signals help doctors to identify illnesses and diseases. Here are some of the machines that detect nerve signals.

• The electrocardiograph (ECG) machine measures electrical signals from the nerve muscles of the heart. It shows whether the heart is beating properly.

Monitoring a patient's heart with an ECG machine.

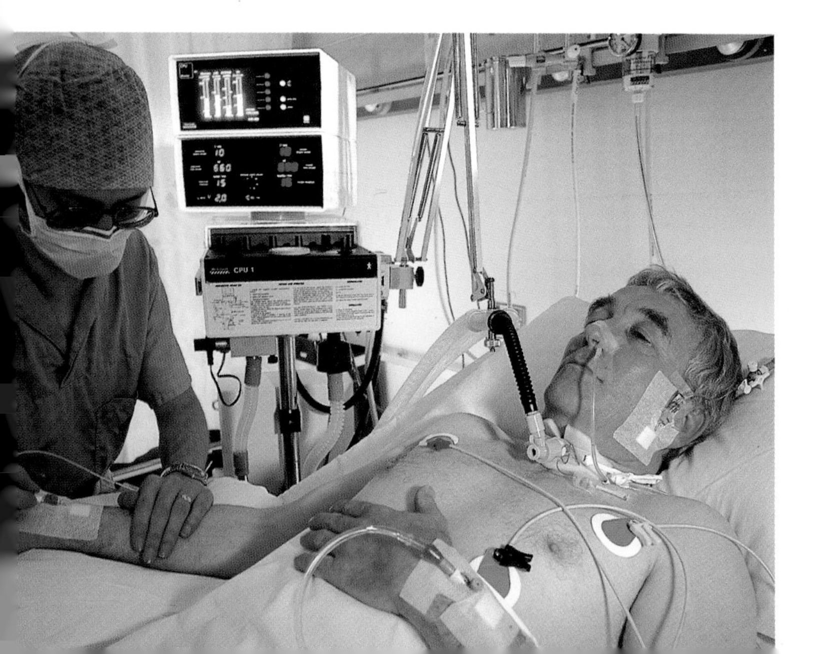

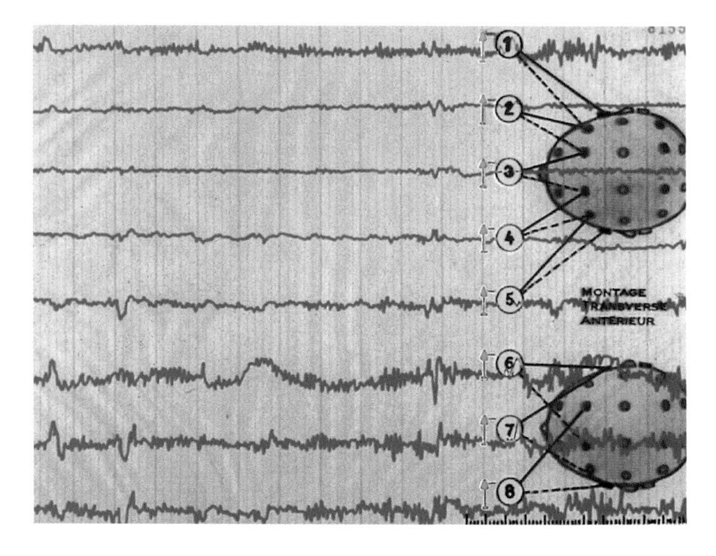

A print-out from an EEG machine shows electrical signals from a brain.

• The electroencephalograph (EEG) machine measures electrical signals from the brain. It shows whether the brain is working normally, or whether it has been affected by a fit or a stroke.

• The electromyograph (EMG) machine measures electrical signals from muscles. It shows if muscles are working properly, or are weak or paralysed.

Pacemakers

The heart is controlled by tiny nerve signals. The signals make the heart's muscles beat regularly to pump blood. Sometimes a person's heart beats jerkily. A pacemaker is a device placed under the skin that sends signals to the heart to make it beat smoothly instead.

FUTURE FILE

CYBER BODIES

The applications of electricity and magnetism are almost limitless. It is almost impossible to guess where they could take us in the future, but here are some ideas to get you thinking:

- Microchips could be put into the eyes and ears of people who have lost their sight or hearing so that they could see and hear electronically.
- A sensor could be fixed to a person's head to detect their thoughts and use them to control a computer or other devices.
- A whole electronic body of microchips and motors could replace a person's body. A person would only need a brain!

MODERN ELECTRONICS

The modern world could not operate without electronic devices. The most useful devices are for communications. They make radio, television and telephones all possible. And email and the Internet would not exist without the electronic computer. The computer is such a useful machine because it can be programmed to do almost any job you can think of, from writing books like this one to flying a fighter jet. Electronics are everywhere – and they are here to stay.

The Internet is made up of millions of computers all over the world.

A virtual reality system uses a computer to create a world that looks and feels real.

Glossary

Alternating current (AC) A currrent that flows first in one direction and then the other.
Atoms The tiny particles that all substances are made from.
Conductor A substance through which electricity flows easily.
Direct current (DC) A current that flows only in one direction.
Domains Tiny parts of a magnetic substance that act like magnets.
Electric charge An amount of electricity. A current is made of flowing charge.
Electrical signal A changing current that represents something, such as a sound wave.
Electrode A piece of material through which electricity flows in and out of a cell or an electrolyte.
Electrolysis Splitting up the chemicals in a liquid using an electric current.
Electrolyte The liquid used in electrolysis or the liquid or paste in a cell.
Electromagnetic effect When a flowing current creates a magnetic field.
Electromagnetic induction When a current flows in a wire that is moving in a magnetic field.
Electromagnetic spectrum The whole family of electromagnetic waves, including light, radio waves, microwaves and X-rays.
Electromagnetic waves Waves made up of vibrating electric and magnetic fields.
Electromotive force (emf) The push that a battery gives to make current flow round a circuit.
Electroplating Using electrolysis to coat an object with a thin layer of metal.
Friction A force that tries to stop two surfaces sliding past each other.
Fuse A thin wire that melts to break a circuit if the current becomes too high.
Generator A device that gives out electric current when it is turned.
Gravity The force that pulls every object in the universe to every other object in the universe.
Hydroelectric power Electricity produced from the energy in flowing water.
Insulator A substance through which electricity does not flow easily.
Ions Atoms that have lost or gained electrons.
Magnetic field The area around a magnet where the magnet's force can be felt.
Microchip An electronic device made from a chip of silicon that contains many microscopic components.
Nuclear reactor A device that creates heat from nuclear reactions (where atoms split apart).
Nucleus The central part of an atom, made of protons and neutrons.
Optical-fibre A thin glass fibre along which light travels.
Ore A material from rocks that we get metals and other chemicals from.
Power The amount of energy used in a certain time.
Resistance The amount a substance tries to stop electricity flowing through it.
Semiconductor A substance that can be both a conductor and an insulator.
Smelting Extracting a metal from its ore.
Static discharge When a charge of static electricity moves from one place to another.
Static electricity The form of electricity that does not flow. Static charges exist on the surfaces of objects.
Superconductor A substance that has no electrical resistance at all.
Transformer A device that changes the voltage and strength of a current.
Transistor An electronic component that can turn a current on and off, like a switch, or amplify a signal.
Wavelength On a wave, the distance between one wave crest and the next.

Further Information

PLACES TO VISIT

Eureka! The Museum for Children
Over 400 hands-on science exhibits, interactive activities and challenges.
Eureka! The Museum for Children, Discovery Road, Halifax, HX1 2NE
www.eureka.org.uk

The Science Museum
Thousands of exhibits and hands-on activities on science and technology.
Science Museum, Exhibition Road, London, SW7 2DD
www.sciencemuseum.org.uk

BOOKS TO READ

Hands on Science: Electricity and Magnets by Sarah Angliss (Kingfisher, 2001)
Science Answers: Electricity /Magnetism: From Pole to Pole by Chris Cooper (Heinemann, 2004)
Science World: Electricity and Magnetism by Kathryn Whyman (Franklin Watts, 2003)
Science Files: Electricity and Magnetism by Steve Parker (Heinemann, 2004)

WEBSITES

http://skydiary.com/kids/lightning.html
In-depth explanation of lightning, safety hints and photographs

http://solstice.crest.org
American Centre for Renewable Energy and Alternative Technology, with information on many different renewable ways of generating electricity

http://science.howstuffworks.com/cat-scan1.htm
How CAT scanners and other body scanners work

http://www.exploratorium.edu/snacks/stripped_down_motor.html
Learn how to make a very simple electric motor

Index

The Earth's Magnetic Field

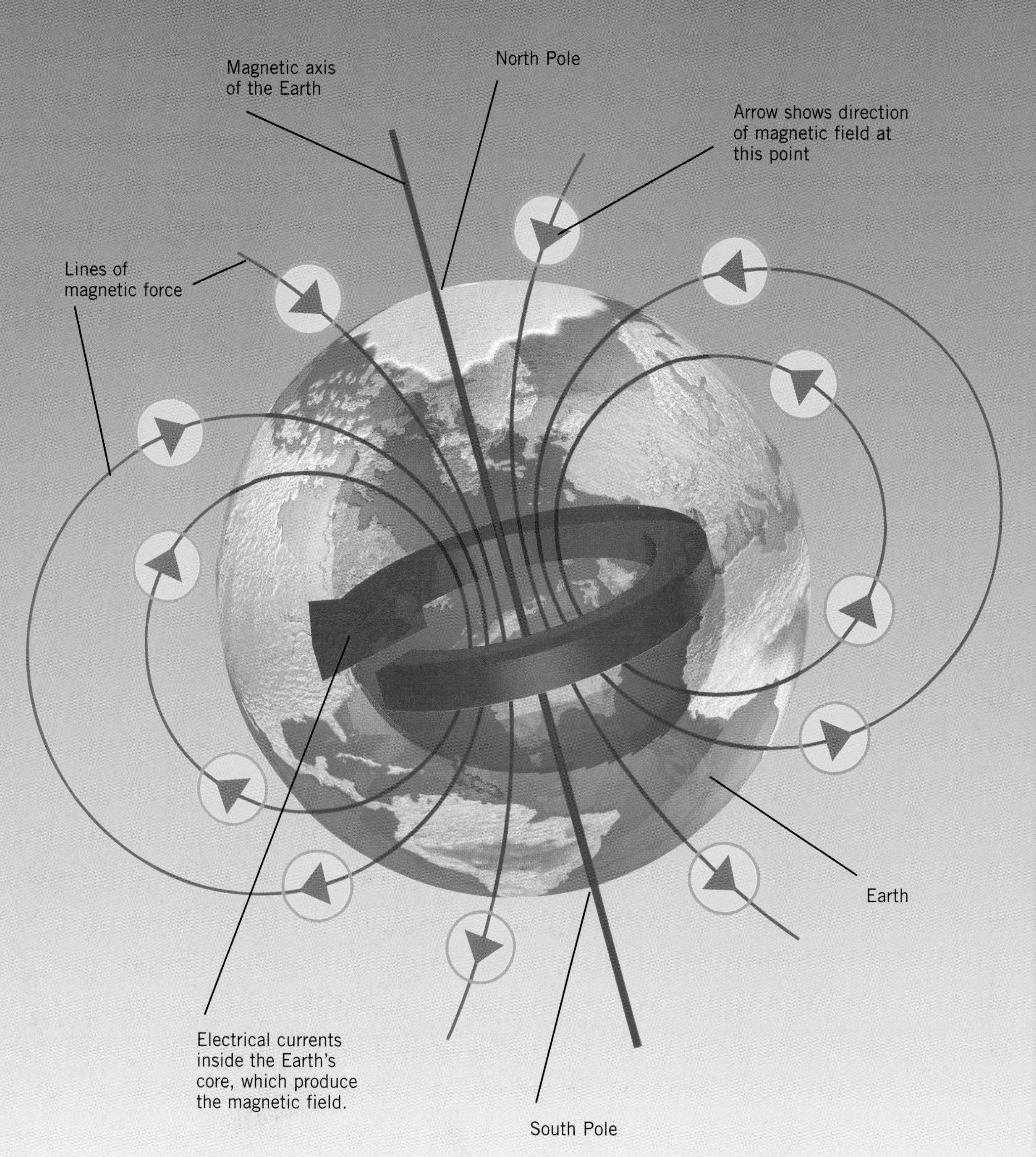